Ray stielo walks upright, proud and over confident across the bridge of the Ravanger, the newest ship built for the resistance on Rihese. Ray summons his lieutenant to his side. Lieutenant Rance he says softely,"We have a code blue aboard this ship". Lt.Rance looks square into his commanders eyes and states" I'll assemble the men at once sir". Lt. Rance grabs the intercom mic and issues the orders that all men must report to the bridge immediately. Private Leo Swifty stands bold at attention awaiting for his commander to speak. Ray walks down from his post and sighs."Men we have a code blue aboard this ship". The men gasp. "A code blue states we have intruders aboard, me and Lt. Rance will interview each inviduail until we find the intruders aboard, that is all". Ray salutes his men while they salute back. Rance states the men are dismissed and they return to their post. "Rance I'm going to my quarters for the night". Ray sighs and head down the corridor to his room, he opens the door and heads to his bed. Rance heads back to his post upon the bridge of the ravanger, takes the intercom mic and asks for the nightly reports. Private Blueie brings him the reports. Their eyes meet and at that second there was something, Rance smiled and took the reports and said "Thank you private". She smiled back and said "No problem sir". Rance turns around and heads back to his chair aboard the bridge. Private Leo Swifty was down at his post in the engine room, working on boasting the frontage power to decrease the amount of fuel the engines use. Frustrated, Leo puts his wrench down to take a break and heads to the day room for some food. Leo grabs some food and sits at a table in the far side of the day room. He pulls out the blueprints for the new engine designs and starts to study them once again. Then he noticed something. He thought to himself for a minute, could there really be a flaw in the new plans. He checks over it again and again, and came up with the same results; the new

engine won't decrease the amount of fuel intake, because the new jet thrusters will cause a backfire. He at once runs down the corridor and heads to the bridge. Once arriving at the bridge he finds Lt. Rance and says." Sir I've been working on the new engines and I think I found a flaw". Lt. Rance looked puzzled and asked him to explain himself. Leo went through what he just found out and Rance looked quite surprised."I'll take this to the commander in the morning. But for now Private you should draw up new plans for the new engines and get back to me immidetiely.

Planet: Nyrae (sister planet to Rihese)

General Shaw wakes up disturbed by a loud knocking at his door. He signs and asks them to wait a minute for him to get decent and heads toward the door. He opens the door and asks" Why have you awoken me this early". He looks for the insignia on the trooper's uniform and states "Sergeant". The Sergeant gasps and says" General Shaw". Out of breathe. Shaw tells the Sergeant to take a second to catch his breath then speak. "General Shaw we need you at headquarters immediately. The General gestures for the sergeant to grab his coat as they head towards the shuttle. Shaw looks at the newly ranked sergeant and asks him what this is about; the sergeant says he doesn't know. Shaw's private shuttle gets to headquarters and gets out and tells the sergeant to give him his coat. Shaw puts on the coat and heads toward the headquarters door. He puts his hand on the scanner and his eye for the rectangle scanner. The doors open and he heads down the dark painted corridor into the control room. When he gets there his eyes are met with the head of state." Head of State Corriban, what do I owe the pleasure". The head of state stares down Shaw and says " Threes been an accident". "What do you mean there's be an accident" Shaw says firmly. Corriban replies "Commander Toey has been assinated".

Shaw is star struck by those words."What, by whom? " We are not sure yet general but we are working on finding out"." But as of right now you are acting Commander and until we have a ceremony, this will have to do". Corriban pulls out two bars and pins them on Shaw's uniform." Congratulations Commander". Corriban states. Shaw doesn't hesitate to speak, "Head of State Corriban no disrespect but do not congratulate me on another man's death". Shaw quickly heads out the room and heads to his private office to grieve for his friend. An hour later his Lieutenant knocks on his door" Sir we need your orders" he clearly states. Shaw takes a breath and says "Get in contact with communications and find out who did this". " Oh and Lieutenant Erly ask the mess hall to send down some of its finest". " As you wish" states Erly. Erly heads out of his Commanders room and heads down to the mess hall to put in the commanders order. " Braylen send the commander you're finest". " Yes sir" Braylen sighs. Erly heads out of the mess hall and down to the elevator. He steps into the four sided titanium elevator and looks for the slot to put his key. " Where would you like to go Lieutenant Erly?" Elry smiles as always the voice reminds him of a board game he played once as a kid. "Commutations". He replies. As the titanium elevator descends, Erly is once again brought back to his childhood. He remembers running through the wind thinking he was so fast that his drunken father couldn't catch him. The doors open and Erly heads down the corridor to a room that is off to the right and opens the door. The men salute him and he the same. Commucation specialist Ride approaches him and says " Send my regards to the commander". " Will do". Erly states clearly. Erly takes a minute to think and then says " I have an important job for you men, and that is to find out who committed the assination". Ride stares into his superiors' eyes and says " We will report to you as soon as we find out sir". They salute each other and Erly walks out of the room.

Outskirts' of Rihese

Commander Stielo awoke that morning from a peaceful sleep, so he was ready to start his interview of the men. But before that Stielo headed for a shower and a shave. Afterwards Stielo heads to the com to call his family. It the coms rang several times before a familiar voice answers. "This is the Stielo residency" the familiar women replied. " Hello Marsha" Stielo replied. " Hey honey how's the new ship treating you?" Marsha says. Stielo quickly replied " Good, how's Caroline doing?". She sighs before replying. " She's, you know Caroline". Threes a knock at the commander's door. " One Second honey" Stielo says. Stielo puts the com on hold and answers the door. Lieutenant Rance was there will the documents that Private Swiftly had found. He glanced over them then returns to the com. " Honey, I'm sorry but I have to go, I love you both". He says. Marsha had an uneasy grin and said she loved him as they hung up the com. Stielo hid the fact that he was interrupted by his lieutenant, and head out his quarters onto the corridors towards the bridge. Rance followed his commander proud. When they got to the bridge of the Ravanger Private Blueie was waiting with the daily reports. She handed them to her commander and smiled towards Rance direction .Rance tried not to smile back but it was too late. He quickly hid it as his commander looks him in the eyes and said "Why wasn't I woken up for this report Lieutenant Rance". Rance replied " Sir I thought it was not of great importance". Stielo tried to hide his anger and simply stated " When any assination happens with a union that we are currently not as peace with, I must be notified immediately". Rance starts to stutter a bit with his words and eventually apologizes. All Stielo said to him was do not let it happen again. Stielo heads to sit in his chair aboard the bridge, when a surprising com comes in. " This is Head of State Rye of Rihese as of right now we are negotiating peace talks with Head of State Corriban and I would like Commander Stielo to be at our next council meeting". Stielo sighs and calls Private Blueie from his com " Private Blueie sends word to Head of State Rye that I will be at the next council meeting, all I need is the time and date". Blueie responded with " Sir he sent the times with the com we just received". Stielo thought to himself for a second, of course he would do that he muttered under his breath." Bring the times to me Private". " On my way sir". She stated. Blueie headed out of her post in the commucations room and started down the corridor when she saw Lieutenant Rance going towards her. She salatued and smiled. Rance salatued and smiled back at her. Blueie continued down the corridor until she reach the bridge then she headed up the stirs toward her commander chair and saluted and handed him the times sent by Head of State Rye. Stielo saluted back and grab the reports from Private Blueie and sent her back to her post. Stielo put the times in his calendar on his pad and noted that he wear his best blues. Rance headed back the bridge and straight to his commander and handed him the reports on which Private Leo wrote. Stielo looked over them several times and then issued for his lieutenant to com Private Leo up to the bridge. Leo was working on the engines when he got the com to come to the bridge, so he put down his wrench and wash his hands, grabbed his coat and head toward the bridge. Leo finally reached the bridge and Commander Stielo and Lieutenant Rance were waiting for him. Leo saluted both officers and waited for them to speak. Stielo and Rance saluted back. Stielo sternly says " Are these accurate Private?" Leo firmly reply's " Yes sir, I went over them several times and had Lieutenant Rance look over them." Stielo looks toward Rance and says " Is that true

Lieutenant?" Rance replies " Yes sir it is." Stielo looks back at the private and says "
When can you start fixing the problem?" Leo takes a minute then says " Now sir." Stielo
orders the private some help with the task and then sends him of his way. Leo heads from
the bridge back down to his post in the engine room with his order help and starts to fix
the jet thrusters first. After Private Leo leaves Stielo turns toward Rance and says
"Lieutenant Makes sure my dress blue are ready for the council meeting." Rance says "
Yes sir". Then he heads down to the commanders room.

Nyrae Headquarters

Back at Nyrae headquarters Shaw was grieving for his friend. Shaw called Toey family
and states to them what had happen and that they are using every resource to find out
who and done this. Afterwards Shaw was interuptted by Head of State Corriban. " Yes
Head of State" Shaw says bluntly. " General Shaw I would like you to attend the next
counsel meeting". " I'll be there just leave me the time and date before you leave today".
Corriban heads out of the Generals room to leave him to griave by himself. Corriban
leave Nyrae headquarters and heads toward his shuttle to take him back to his office.
Corriban arrives at his office and asks his assiatant to send word to Commander Talee for
a meeting. Commander Talee arrives at Head of State Corriban office and is moteined to
head into his office and sit down. Corriban takes a second to think before he speaks and
says " Commander Talee you have an excently record with the Core." Talee stares
blankly and says " Thank you sir". Corriban takes another second to think and then says "
I have a very important task for you Commander". She looks blankly again " Yes sir what
is it?" Corriban stares back and states " I need your team to do some reconizance." "
Where sir?" she says. "Rihese". He states. She gasped " I'm not no to question orders but

why sir?" He smiled grimly " Beacause I beleive they assinated Commander Toey early this week". She was puzzled. " This meeting never toke place and this conversation did'nt ether, understand?" " Yes sir I'll asemble my men immedately sir". Talee gets up and heads out of Head of State Corriban's office and heads to plan the next task ahead of her. Corriban erased the voice recorder under his desk of the last twenty minutes of his and Commander Talee meeting. Corriban picks his mic up and asks his assiatant to get his best dress suit ready for the counsel meeting.

Planet: Tel-Jihad Head of State Rayal office

Head of State Rayal awaits quietly in her office for Grand Monarch Sion to return from his current mission in the outer world. Rayal turns off her recorder at the end of her desk and waits for him to arrive. Rayal office door opens and a dark tall are very mascuclen man steps forward. Rayal greets him with a smile and motions him to sit down. Sion sits down and starts to speak. " My current task in the outer world went very well Head of State Rayal." " Very good Sion, you are a excellent soilder for the Tel- Jihad." she says softly. Sion then motion towards his briefcase and pulls out the reports from his mission on the outer world. Rayal take them and looks over them cautiously, makeing sure every detail was perfect. She then looks at the Grand Monarch proud and says " Has the next task been started Sion?". He smile and with a grin says " We have started production, we should be ready very soon Head of State Rayal." She smirks and says" Well done, Now I have to leave to go to the counsel meeting soon, and I would like you to join me in this event." He hestates for a slipt second and says " I'm a soilder, not a politician, But only because you asked I will be present just send me the time and date and I'll have my best uniform ready". " I look forward to seeing you there, I'll send the time and date to your pad, you are dissmissd Sion". Sion gets up and statues his head of state, and heads out the

her office. After Sion left Rayal made sure the recorder didn't record anything of the

meeting and then turned it back online. Rayal then checks the reports once again and then

sureds the evidence. Rayal then com her secertary into her room to take the evidence

away,far away so she could never be held accountable. Sion heads toward his private

shuttle to take him back to his briefing room. As Sion approachs the black basanol

building, he turns around for a quick second to make sure nobody is following him. Sion

puts his hand on the for body scan, it scan perfect as it always does. The door opens and

Sion heads to the elevator, the elevator doors open and Sion looks for the dashboard to

key in his password to go to the basement bunker.The elevator reaches the bunker and

Sion takes a step out. Sion heads to the farthest door to the left and scan his hand again.

The finally door opens and he is over taken by cheers of joy by his fellow comrads. Sion

smirks and statues his men and they statue back. " Men issue the order to begain

production for the new ships." Sion smiles and find his chair as it always is in the middle

of the rooms floor, he sits and takes the reports from his lieutenant. Sion looks at his

lieutenant and hands him some paper work and says " Lieutenant Connor make sure these

documents get destoryed." Connor stares into Sions eyes then noticed his mark of honor

the black sun tatooed above his eye. " At once Grand Monarch Sion". Connor takes the

documents to the room on the right of the farthest corner and proceeds to destroy them.

Sion issue's a couple more orders for the night, then waits for his lieutenant to return with

good news. Connor heads out of the room and back into the comand room where he gives

his superior the news of the documents being destoryed. Sion rested assurd that his

lieutenant compeleted the order given to him, he tells him he needs his rest for the night for tomorow is another big day.

Planet : Tokeal Counsel Meeting

All pleasant please rise and anounce yourself and your companion the hooded figure

speaks. Head of State Rayal with Grand Monarch Sion. Head of State Rye with

Commander Stielo. Head of State Corriban and General Shaw. He droped his hood down

by his side and says, " Now I will anounce myself Head of the Union Rylan and

accompanying me are my bodyguards Norton and Ture."Rylan sighs then says " Head of

State Rye you can begain." Rye ruffles some papers and get ready to speak when Stielo

whispers something in his ear, which makes Rye stop before he even starts. Rye pauses

and looks into Stielo eyes then he turns back toward the counsel members and states "

We of the resistance are contiuning are peace talks with Head of State Corriban starting

with this, We would like to become partners in trade and the miltary under certain

circumstances of course, but thats for a later date." Rye then motions for Corriban to

speak. Corriban all ready had his documents ready for his speach. " Head of State Rye we

are currently investigating the murder of are beloved Commander Toey, and until we

solve this all peace talks are going to have to be turned down, for the time being".

Corriban turns toward Head of State Rayal and motions for her to speak. " We have no

objects here" she said with a smirk. Head of the Union turns toward Head of State Rye

and ask him " Do you have any objects?" Rye waves his head no. " Ok then this Cousel

meeting is adjured" the Head of the Union states to everyone. Rye turned quickly to

Stielo and told him to meet him in his office as soon as they get back to Rihese. Stielo

noded in agreement and then they both headed toward there shuttle. Corriban and Shaw

walk out of the meeting chamber unto the hall and start talking as they walk to there shuttle " You handled that very well sir" Shaw gestured. Corriban look at him with a proud look and said " Thank you General". Have you had any word from commucations about are recent death?" Shaw pulled out his pad and said " Just got the report sir". Corriban with a smile on his face said " Good work well discuss this at my office, when we get back to Nyrae". They finally reach there shuttle and head of into the stars. As Sion and Rayal reached there ship they both turned for a quick second and made sure nobody had followed them. They board the ship and quickly shut the door. Rayal look at Sion and said " Hows the production going Grand Monarch Sion?" Sion reply's " Almost ready sir". Rayal smirks and says " Well done Sion, well done". There ship quickly takes off into the stars.

Planet: Rihese Head of State Rye's office

Both Rye and Stielo both enter Rye office at the same time and toke there seats. Rye leads off with " Well that went well". Stielo for some reason couldn't understand his sarcatism." Head of State Rye I think we might have a new problem very soon with Head of State Corriban". Stielo states very seriously. Rye takes a moment to think and then says " I firmly think you are right, but we have no proof, so until then do standard produce Commander". Stielo nods then stands up and salutes his Head of State and heads for the door when Rye ask him " Oh and Commander I suggest you start interview for men aboard your ship immeitdary". Stielo nods once again and heads out of Rye's office. One hour later Stielo arrives aboard the Ravanger and beckons his lieutenant over to him. " Lieutenant Rance prepare documents to interview the men starting with Private Leo Swifty". Rance salutes and heads toward the processing center to get all the documents ready for the interviews. Rance then heads back to Stielo and greets him with a salute,

Stielo salutes back and says " Is everything ready Lieutenant Rance?" Rance looks at him and reply's " Yes sir, everything is ready, when ever you want to start I can have the men one by one up on the bridge". " No not on the bridge in my private quarters" Stielo reply's." Go ahead and send word to Private Leo to meet me in my quarters". Stielo heads toward his private quarters and reads over the documents of private Leo Swifty as he walks. Leo was in the middle of finshing the new jet thrusters when his com rang, so he answers it "Private Leo here". " Yes this is Lieutenant Rance speaking, you are at once to report to Commander Stielo's private quarters". Leo sighs " I will report at once sir". Leo stands up and the only thought that runs though his mind is that they were starting the interviews to find the intruders aboard the ship. Leo arrives at Stielos office but before he could open the door, Stielo opend it for him " Private Leo come in and sit down". he said. Leo steps forward and sits down in the chair across from his commander. Stielo looks at Leo in the face then looks back at the documents that Rance gave him. He sets the documents aside and begins to speak " So Leo, I can call you Leo right?" " Yes sir". Leo gestured. " So you have been with the Resistance for a while now right?" " Six years this March sir". Leo replied. " I was looking into your file, you have done some outstanding thing in the engine department of each commison you have had". Leo sighs " Sir I know this is an interview but I can assure you I'm not the interuder aboard your ship." Steilo stares Leo down for several minutes and says "How the work on the jet thrusters going along?" Leo looked puzzled, " Well sir I'm almost finshed, then I will move onto the main engines". Stielo simled and stated " Very well you are free to go Private." Both men salute each other and Leo heads toward the door when Steilo says " Very good work on those thrusters, keep it up and you might be up for a promotion soon". Leo smiled at

Stielo and thanked him.

Outworld of Rihese

Commander Talee and her men head toward the exit of the passenger ship The Freedom. As she heads for customs she gets her pad so the custom agent can scan it. She smiles at him and it passed, she then heads for the meeting point in an abandon building not to far from the starport. So she grabs a shuttle and tell the driver the address, he looked at her kinda puzzled and then toke off. When she arrive at the abandon dark blue building it was ruin, glass was shatter everywhere and wall where starting to cavein. She paid the driver and grab hers bags out of the back of the shuttle and headed toward the door of the building, which of eye sight was the only thing intated that she could see. She pulled out for pad and scan it so the door would open and it did. She was the first of her men to reach the meeting point, so she decided after a long journey from Nyrae that she could go down to the local pub and grab a couple drinks before the rest of the men check in. So she sprayed some of her best perfume on to cover the shuttle smell, and headed out to signal a shuttle to take her to the pub. Once she reached the pub, she payed the driver and got out and head to the pub's door. She noticed a sign and it read No Weapons of any kind, its a good thing she left her gun at the meeting point she thought. The pub door opened and she found herself at the bar stool next to the bar. She gestures the bar keep to send her way his best drink and he does. After a couple drinks she turned to look around and found in the corner a attractive young solider. She looked at him and smiled and he smiled back. He gestures for her to come over and so she does. As she stumbled across the pub's floor she finally made it to the corner where he was sitting. She looks at him and says " Whats your name cutie" The young man blushes and says " Private Curtis". The two shared drinks and carried on for a couple of hours. Talee decides she has spent enough time

enough in the pub and gestures for the door, so they hug and she heads out back to the

meeting point, staggering as she waves down a shuttle to take her back there. She arrived

at the abandon building once again only to find that her men have arrived, so she briefed

them and said she was going to head to bed for night and they will start recon in the

morning.

Planet: Nyrae Head of State Corriban's office

Corriban entered his office first and sat down to turn off the voice recorder under his

desk, Shaw soon followed but sat in front of him. Once the the recorder was off Corriban

made the motion for Shaw to bring out the reports." Head of State Corriban my

communcations team have reason to beleive that it was in fact." He was cut off by a

lound knocking at Corribans office room door. Corriban angrly yelled " What is it?" Pale

face Sergant frowned and said " Sir we have a problem". Corriban quickly replied back

and gestured for the Sergant to come inside his office and shut the door " Well get to the

point". The young Sergant stumbles to shut the door and says " Sir there's been another

assination". Corriban looks at Shaw puzzled and said " Who"? The Sergant hands

Corriban the reports, and Corriban reads over them, Not even a second later the Sergant

pulls a knife out from his belt and holds Shaw hostage against the office door. Shaw

looks at Corriban and Corriban looks back but with a grim smile, as he pulls out a gun

and shoot towards Shaw direction. Shaw hits the sergant with an elbow toward his ribs

which causes him to bend over in pain, while the shot hit him in the head. Corriban looks

for Shaw out of the smoke from the gun shot as several bodygaurds and assistants enter

Corriban office locked and loaded ready for a fire fight. They quickly grab Corriban and take him out of the room. Shaw gets up of the floor after all of Corriban's men leave and still finds the dead sergants body on the floor by the foot of the door. So he coms some of his men and tells them to take the sergant aboard his ship. Shaw decided that night that he must return to his ship and take this into his own hands, especial since they threaten his life, so he toke the reports on Corribans; desk head back to his shuttle. When Shaw finally return to his ship that night his men had all ready brought the dead sergant and cross matched records to find out who it was really. Lieutenant Erly was awaiting for his general for several hours , when he finally arrived he handed Shaw his reports of his assissan. Shaw puzzled at first and then read the reports over again to make sure, he read it right then he looked at his lieutenant and said " So let me get this straight you have no idea who did this?" Shaw said angrly. The lieutenant shiver for a second before replying back to his general " Yes Sir, But". " But?" Shaw rudly cuts him off, before Erly could finsh his sentence. Erly gulps. Shaw stares down his lieutenant before speaking again " Send the orders for mass production of the new Bisflak 49's, I want every ship equiped with them immidanitly". " Yes sir" Erly stops shivering and salutes his general and heads off to file the orders.

Planet: Nyrae Corriban's Private Bunker

Corriban finally stands up from his desk in the underground air raid bunker and signals his nearby bodygaurd to his attention. " Contact General Shaw immediarty and ask him if he has found out who plan this attack". The bodygaurd runs to a com on the desk and dials out toward Shaw's ship in orbit." This is Corriban's bodygaurd Tote, looking for General Shaw". he says. The com dispatcher answer " I'll direct you to him". Shaw was on the bridge of his ship when he got the com. " This is General Shaw, Hows Head of

State Corriban?" Tote answered back " He's safe for the time being, General, Head of State Corriban ask me to ask you if you have and leads on this new attempt of assination?" Shaw sighs " No we don't but tell Head of State Corriban be are being production on the new starfighters immdeiatry". Tote puts Shaw on hold and tells Corriban about the new production and Corribans reply's " Tell General Shaw to find out who did this and make them pay". Tote nods understandingly and heads back to the com to tell Shaw the news " General Shaw, Head of State Corriban says find out who did this and make them pay." Shaw signals his office to make new orders to be sent out to the men before replying to Tote." Tell the Head of State It's all ready been done, oh and Mr. Tote keep Corriban safe." Tote assures Shaw that Corriban will be safe then turns off the com and heads back to his Head of State. Corriban ruffles the reports given to him from General Shaw and begins to read them throughly. He comes across some starting evidence that The Resistance may be behind all the assination of Commander Toey and the recent attempt on him and General Shaw. But Corribans knows that no matter how much he might have a feeling it was The Resistance, it's all about the evidence behind it. So Corriban desides do some through research to find who is behind these attempts. After Shaw sent the orders down the chain of command all personal abroad his ship we well awhere of the new production, one of these orders were for the pilots to begin silmutioning the flight produces of the new starfighters. Meanwhile Shaw was up to design plans to a new ship for his navy, one that couldn't be destoryed by regular blaster cannons, one that would strike fear into the oposing side. Once Shaw left the bridge to work on the new plans, Lieutenant Erly heading to his private quarters to make a com call. He dialed out,"This is Lieutenant Erly, and I need to speak with Commander Talee".

he stated. " Sorry Lieutenant but Commander Talee is on leave right now". the dispacher replied. Erly thought to himself for s slipt second " On leave when we are at a crisis, thats blastphemy". The was a pause between the two. Then the dispacher replied " Lieutenant I can take a message and send it to Commander Talee as soon as she returns form leave." Erly muttered under his breathe." No thank you". He hung the com up and headed back to the bridge to issue more orders for the night, then preceeded to head to bed and put Lieutenant Pasis in charge.

Outskirts' of Rihese

Stielo awoke the next morning to reports of the assnation attempts on General Shaw and Head of State Corriban. So he sent the reports to Head of State Rye office with a worried messge of being on the brink of war. After he sent the reports Stielo headed towards the day room to get some food. Stielo saluted Private Braylon and he saluted back, then he grab his food and headed for a seat in the back on the day room. Braylon noticed that his Commander was worried about something so he headed over to him to find out. " Sir, are you ok, you seem worried". Braylon replied. Stielo thought his words wise before speakign so he didnt alarm the men of his recent worries of war. " I'm fine Private just have some stuff going on back home". Braylon not buying his story assures his Commander that if he needed anything he would be there for him and headed back to make some for food for the next meal of the day. Stielo finshed his food and headed back up to the bridge to await Head of State Rye's reply to his messge he sent earlyer this morning. When he got to the bridge he notice his pad had a message on it from his Head of State so he headed to his priavte quarters to read it. The message read *Commander Steilo I recieved your recent message and I am worried about the brink of war being brought upon also, espesilly on something we didn't do, So we will follow standard procedure doing this time, and you are not to alrm the men of are recent converstation. The message with sincerly Head of State Rye.* Stielo erased the messge so nothing could lead back to him and decided it was time for the next interview so he commed for Lieutenant Rance to send up Private Braylon, so he can begin questioning. Braylon got the com from Lieutenant Rance and procedeed to head up to his Commander quarters and with he brung some of his finest alcohol. When he arrived the door was open so he stepped in only for Stielo to greet him, both men saluted and the interview started. " Thank you for the alcohol Private, I'm sure i'll drink in the later dates to come". Braylon smiled and noded in aprrectation. Stielo contuined " I looked though your file and found you have only been with us for two years Private, so you understand why we might what to question you". Braylon confused stated " Sir I left my wife and kids to join your navy, to sever are home planet from all ememies, foreign and domestic, so when you question my love for this navy I take that offensively". Stielo smiled knowing Braylon wasn't the intruder abroad his ship and responeded with " Private Braylon, I do not trust easyly but I also do not think you ae the interuder aboard this ship, but this is standard prodocall". Braylon nodded understandly that STielo didn't mean to insult his love for the navy." But I have one more question for you before you are dismissed". Braylon a little worried answered " Yes Sir?" Stielo pulls out some documents and handed them to Braylon. Braylon read over the documents over and over and then begain with " Sir, you want me to spy on are men in the day room?" Steilo quickly replied " Not spy just get to know the men a little better, and you will report back to me everyday". Braylon not happy about his new orders decided to agree to them. Stielo dismissed Braylon back to his post in the day room, and headed back up to the bridge to get the rest of the morning reports from Lieutenant Rance.

Planet: Rihese Commander Talee's special recon unit

Commander Talee begin the morning issueing recon assignments to the men. She headed out the abandon building and waved down a shuttle and asked the driver to take her to the Public records buildings downtown, the driver nodded and they were on there way.When she arrive at the building she quickly heading inside and ask the clerk where the pad was to find out all the history of The Resistance navy. The clerk showed assited in showing and left her on her way. Talee quickly searched out all teh recent miltary assignments and all she found was just simple fleet movments around the planets orbit. She was puzzled so she searched several other ideas that came to mind and still found didn't find anything. After hours of desprete search she decided it was time to call it quits and head back to the abandon building hopeing that maybe some of the other men in her unit found something. When she arrive the men didn't salute because they were undercover so she ask for the reports.Talee had taken the reports given to her by her sergant and headed back room where she begin to read them over and over looking for some evidence to point the assnation of Commander Toey against The Resistance. After twenty minutes and still fund nothing she decided enough was enough and pulled out her pad and message Head of State Corriban about the news so far. After she message Corriban she decided to give it one last try and headed back to public records, this time she knew right where to go, so she seated herself and got right to work.The only documents found between the two planets where of peace talks so she got an idea and proceeded to head down to the miltary headquarters.When she arrived she was greeted and a Lieutenant. " Hello my name is

Lieutenant Raloud, How may I help you?" " Well I was wondering if you think these peace talks are going to work out?" She stated. The lieutenant looked at her and usherd her inside the building." My office is on the right down the hall please take a seat and I will be with you shortly". And She did. Meanwhile the lieutenant headed toward the commuations room in the building and decided he didn't have a good feeling about so he toke her picture from the sercuity camera and ran her fac in the civillian database. He found nothing, so puzzled he ran it thought the interglalatic data base and found that she was a commander for the Nyrae. He quickly commed Head of State Rye." Head of State Rye here." " Yes sir we a commander form teh Nyrae asking questions about peace talks". Rye sop in his tracks for a slipt second." Well escort her to my office". he replied. The lieutenant accpeted his orders and started to head back into his office. While the lieutenant was gone Talee hacked his computer and still didn't find anything about assiantion orders, she then heard him talking down the hall so she quickly jumped back in her seat and awaited for him to arrive." Mrs. are Head of State would like to see you, so i wil be escorting you there." She smiled and followed him. They arrive at the Head of State's building. Talee is escorted right into Rye's office." Please come in and sit down Commancer Talee". Rye gestured. Talee sat down." I kinda figured you ran my face in the database". Rye smiled. " Not everyone asks about the peace talks around here". He replied. "Well to get right to the point, we did not assinate Commander Toey, and I know that why you are here doing recon". She was puzzled to know that he knew that." This my planet I think I would know if a commander and her men would come here." She sighed." Head of State Rye I we had no intentation of starting trouble between the two planets". " I understand that I just wished you would have came to me first". he replied. "

Now I am going to have my men escort you and your men back to Nyrae immidiarty".
She nodded understandly and was escorted out his office. Rye pulled out his pad and
messaged Commander Stielo of the news of the recent Commander Talee's mission.

Planet: Tel-Jihad Head of State Rayal's office

Rayal hands Sion the lastest documents his commucations team received." Now Grand
Monarch Sion have you heard the news about the tentation between the Nyrae and
Rihese". Sion pateintly reads over the documents and says " Yes Head of State I have." "
Good and how's the production going,for we might have a war upon soon". Rayal stated.
Sion smiled and with a big grin said " We are almost ready". " Good send word the all the
commanders in the navy, for war is on the brink". Sion saluted his head of state and left
his office, and headed back to his ship to give the orders to all the commanders. Sion
arrived at his ship only to be greet by his Lieutenant. " Lieutenant I need reports from all
the commanders and production reports of the new ships". Connor saluted and headed to
do as he was ordered. Sion couldn't help but to smile about the new news, about the soon
to come war, but he wondered which side he must take.

Outskirts' of Rihese

Stielo recieved the new message form Rye and read it couple of times before replying.
*Head of State Rye I read your recent message about Commander Talee asking questions
about peace talks, and the assination of Commander Toey, I also have news of anotehr
assiantion attempt but on the lifes of General Shaw and Head of State Corriban, So I fear
are worst unsumptions might come true. Sincerly Commander Stielo.* After Stielo sent
the message form his pad, he issued some orders he hasn't issue ever in his career. He

called his Lieutenant to his attention and told him to send warning to all commanders in the fleet of an attack, by and unknown enemy. After issueing the those orders to Lietuenant Rance Stielo rose from his chair commed Private Leo to the bridge. When Leo reach the bridge he gave the new reports about the jet thrusters being done and that he had almost compeleted the new engine. Stielo read the reports and thanked Leo for his good work. Stielo returned to his seat at the helm of the bridge and motion for one the the sergant's under his command. Both men salute each other before Stielo speaks. " Sergant issue my next order carefully" The sergant was puzzled. " Yes sir" he said. Stielo rubbed his head before speaking. " Send a full patrol to Head of State Rye office, where he goes you go, understand". The sergant nodded and Stielo dismissed him. Stielo decided to call it a night so he headed back to his quarters to com his wife. " Hey Marsha Stielo speaking". she said happly. " Honey do you remeber the private com channel I used awhile ago from the house?". Marsha thought to herself for a moment. " Yes love I do". " Good com me back on it". Stielo waited for his wife to com him back only to tell her about the news that war may be upon The Resistance soon. The com rang Stielo quickly answered it with a smile " Honey, what I'm about to tell you know one can know". She sighed with a worry then agreed." We are on the brink of war with the Nyrae, and I need you to take youself and Caroline and leave Rihese, head to Tokeal I'll have someone waiting for you there. " She wonder why but didn't ask " I will love". " Oh and Honey don't stop till you get there". Stielo said with a sigh and told his wife he loved and will be with them soon. Stielo then hung the om up and made a couple more coms before heading to bed.

Planet : Tokeal Head of Union office

Rylan was reading over the recent news about the assnation attempts on Nyrae when there was a knock at his door. Rylan commed his assistant to let the person in. As the door open, Rylan noticed the person was hooded, so he couldn't clearly see his face. The hooded figure sat door across from Rylan. Rylan confused asked " How can I help you?" The hooded figure sat in the seat for a couple of minutes before he spoke. " Well Head of the Union with your death". Before Rylan could alarm his bodygruards he was dead. Then the hooded figure left the room and vanished.

Planet: Nyrae

Commander Talee left the shuttle, The Freedom and headed towards Head of State Corriban's office. We she arrived there she was greeted by a armed patrol outside. She handed the officer in charge her papers and was let inside. When Talee got inside the building the assitant ushered her into Corriban's office. Corriban was reading the news and reports given to him by General Shaw. Talee saluted her Head of State and sat down. Corriban finshed reading his reports before speaking to her. " I got your report." He said bluntly. " Yes sir". she said softly. Corriban rubbed his eyes before speaking again. " Well have you heard the news about Head of the Union Rylan?" She was puzzled. " No I did not sir". Corriban handed her the reports on Rylan, and she read them. She stopped with it when it got to the part about the unknown killer and she rose her head to look at Corriban again. " You don't think The Resistance did this, do you?" she said. Corriban stared back at her, then ruffled thought his draws of this desk and handed her another document. She begin reading a second document until she finshed it. " Commander Stielo issued a patrol for Head of State Rye right before the Head of the Union was assinated". Corriban said. She thought for a moment. " Sir I still don't think". Corriban interuped her.

" You serve your people right?" He stated. " Yes sir I do." She replied with a proud smile. "Good, then prepare your men for war, and send these documents to General Shaw". She agreed and saluted her Head of State and left. After Talee found a shuttle back to her ship she sent the orders over to General Shaw. She sat aboard her ship when the com came in. " Mam we have General Shaw on hold for you". the private said." Turn the com to my station". she replied. " This is Commander Talee". She stated. " Yes Commander, since you are awhere of are current orders, then I don't think your going to question this one". Shaw stated. Talee was dead still until Shaw spoke again. Shaw with a big grin said " Prepare your men to fire upon The Freedom". Talee was shell shocked before she replied. " Yes Sir". She said with a heart broken sigh. Shaw endend the com. Talee's lieutenant approached her. " Im awaiting for orders Commander". he said. Talee toke a moment to think before she responded. " Make are current course to engage The Freedom and order, the men to prepare to fire upon when we can intersept it". The lieutenant saluted and issued the orders down the chain of command. There was a sudden long pause in Talee heart, for she knew she was about to fire upon a passenger ship.

The Shuttle: The Freedom

The shuttle was just leaving orbit when Trisha knew she forgot something. She sighed " It's too late now". She thought. After a couple hours in orbit Trisha decided it was time to call her parents. " Hello" a older man voice said. Trisha smiled she knew it was her father. " Dad, I'm on my home". she replied " Great I'll tell your mother, and we can't wait to hear about your trip" he replied. Trisha hang the com up and decided it was time to sleep, for she had a long journey back home.

Commander Talee's ship

Talee was sweating in her palms when her lieutenant appraoched her. " Ma'ma we have the shuttle in our range". He said. Talee had a small shake before issue her next order. " Fire when ready Lieutenant". she said with a shake in her voice. Her lieutenant issued the order. The shots were fired and The Freedom was gone. There was a long pause on the bridge as most of the men including Talee saw The Freedom disapear before there very eyes.

Planet: Rihese Head of State Rye's office

Stielo was hurried into Rye office. Rye handed Stielo some reports. Stielo began reading them and thats when Rye notice he had a shocked look upon his face. Stielo read the report once again not believeing them before Rye interrupted him. " Commander Stielo, I do not know what to do about this". Stielo was still looking at the reports, then he took a moment to think before he responded to Rye's statement. " Sir I'm not sure, but your my Head of State and you need to make an order, and fast before the media do for us." Stielo said. Rye eyes were tear jerking after Stielo made that statement to him. He then called his assiatant into his office. " Yes schudule a press meeting as soon as possible." He said. Rye then looked at Stielo and asked if he would join him at the meeting and Syielo noded in agreement. Rye's assitant knock on door before coming inside Rye office. " Sir, there all ready waiting outside for you" his assiatant said. Rye looked at Stielo one last time before speaking. " Lets go get this done". Rye sighed and heading out of his office with Stielo following outside the Head of State's building. When both arrived outside they were swarmed with media and questions. " I'm not taking any questions right now". Rye said firmly. The media stopped dead in the tracks and waiting for Rye to speak again. Rye rubbed his head and begain to speak again. " To the people of Rihese, I am your Head of State, I am your leader, and most of all I am your friend". Rye rubbed his head again flustered with thought. " I am dearly sorry to all the familys that lost loved ones aboard The Freedom". The media was still quiet, busy taking notes of everyword Rye said and he knew that so he chose his words very carefully. He sighed on final time before issue his next order. " People of Rihese we are at war". There was a silence in the crowd for a slipt second before the chattering and swaring begin. Rye quikly quieted them down. " Now I know you must be shocked, but we have tried peace talks with the Nyrae mutiple times, and all they do is dodged us, we have offered resoucres and supplies to help them, and they have pushed us away." The crowd was still quiet when Rye spoke his final words. " We will not let an attack on our people go unpuished." Rye sighed and dismissed the crowd and headed back up to his office with Stielo behind him. Rye quikly let Stielo in before stutting the door behind him. Stielo stood at attention awaiting his order from Head of State Rye. Rye pulled out some blank doucments and wrote the new orders on it. "You are to take this to every Commander in the navy". After Stielo had taken the documents from Rye he knew the meeting was over so he salatued his Head of State and headed to headquarters. Stielo boarded his private shuttle holding the new order Rye have given him, he read over them again and sighed. Once the shuttle reached The Resistance headquarters he quickly rushed inside. The desk clerk greeted and salatued him. " Send these order down the chain of command" Stielo said. The desk noded in approval and went to the communcations room to send the orders. Steilo headed back to his private shuttle to take him back to his ship.

Planet: Nyrae Head of State Corriban's office

Corriban's was sitting at his desk and going over the orders he issued yesterday for the attack on the shuttle The Freedom. When his assitant brought the reports of Rihses declaring war against them. He smiled as he read Head of State Rye's speach to the press and then commed Sion. " This is Head of State Corriban direct me to General Sion". The dispacther repled with " Yes sir". and directed him over to Sion quarters. Sion was going over some production reports when he got the com. " General, how's the production looking?" Sion grinned and said " We are done most of the ships are equiped with the new starfighters, and many more are on the way sir". Corriban laughed. " Good, now prepare your men for battle, Rihese has declared war against us." Sion couldn't help to smirk. " Now General moblize Commander Talee's disivion immideiatry, we need to protect are borders". Sion noded. " And on more thing General, send a patrol to my office". Sion noded once more and then he hung up the com. Sion got up and headed out unto the bridge to issue the new orders. After issuing the new orders Sion sat at teh helm of the bridge and motion for his lieutenant. The lieutenant salatued, before speaking. " Yes sir?" " Com Commander Talee to the helm of the bridge" Sion replied. " At once sir". lieutenant noded and then went to the deck to com her in. "This is the Warhog" the dispatcher replied. " This is General Sion , com me though to Commander Talee" Sion said. " Yes sir". she replied. Talee was in her chair on the bridge when she got the co patch though. " Yes sir?" she said. Sion grinned before replied " You are awhere that we are at war with The Resistance so for your new orders sent from Head of State Corriban take your divsion and defend sector A". " I will be there at once sir". she said. Talee sighed before issueing orders for her disvsion to move to sector A.

Planet: Tokeal Shuttle Bay

Marsha and Caroline left the shuttle and heading to get checked out by customs. The customs agent who was checking the documents was tall and linkie but pale in the face. " Ma'ma can i see your documents?". He asked polietly. " Yes sir". She smiled and handed him the documents. He stamped a red stamp and they preceded to find the person her husband had sent for her. When they reached the cargo bay they grab there bags and headed towards the passenger bay. A tall dark man was holding a sign that said *Marsha Stielo*. She looked at and reconigze the tall dark man as her husbands uncle. " Hello Coren" she gestured. He smiled. " Hello Marsha and Caroline, Now I don't mean to rush you but we need to get going so let me grab your bags." She handed Coren her's and Caroline's bags and they headed toward his private shuttle. Once the shuttle was going and she was comfortable she begin to talk. " So long time Coren, how have you been?" Coren was busy surveign across traffic before he responded to her question. " Good, and how are you guys doing?" She smiled once again before responding. " We are great, do you know where we are going?" " Yes, to a safe house, were noone will find you two, your husband sends his reguards but will be in touch with you soon." He replied. She sighed. The shuttle stopped at a broken down building that was in ruin. Marsha was kinda worried at first being safe in a broken down building, but she trusted her husband and grab Caroline and they headed inside. Once inside Coren put down there bags. " Marsha I have to go run some erins for Ray I'll be back soon, don't open this door for anyone". he stated. " How will you get in?" she explained " I have a key" he said, then Coren headed out the door locking it behind him. Marsha started un packing hers and Caroline's bags, when there was a sudden knock at the door. She ingored it ast first but there was another knocking, so curiously she looked out the peep hole of the door to find a hooded black

figure looking back at her then with a blink of an eye it vanished.

Outskirts' of Rihese

Stielo was too busy checking on status reports and training reports, to notice his

lieutenant was nowhere to found. So he quickly tried to com him, but got no answer,

Stielo decided he would go down to commucations to see if he was there. When he

arrived Rance was not there, so he asked a couple of the men and still got no answer.

Stielo tried to com him again, and he picked up this time. " Lieutenant Rance, where are

you?" He said curiously. Rance jumbled the com for a sec before replying. " Sir I was

out testing the new jet thrusters on the new starfighters". Stielo shugged his shoulders and

told Rance to get back to the bridge. When Rance finally reached the bridge his

commander was waiting for him. " Lietaunt we have orders to move to sector A, prepare

the men for launch." Rance salatued his commander and issued the orders given.

Sector A

Talee was in her private quarters when she got the com that Resistance ships where in the

sector, so she prayed for guidance. She left her quarters in a hurry and headed to the

bridge. When she reach the bridge she commed General Shaw. " Sir Resistance ships are

in sector A I need orders sir". She explained. " Fire at will commander" Shaw said. She

hung up the com and her men were looking at her awaiting her orders. She was puzzled

and confused because she knew they didn't do the assianations. But she toke a deep

breathe. " Men fire at will".

Aboard the Ravenger Sector A

The first shot was fired towards Captain Stoney's ship. " Mayday..Mayday..Commander Stielo this is Captain Stoney we have to..." The com cut off. Stielo was clinching his fist and issueing orders to return fire, when another Resistance ship was lost, this time is was Captain Leug's Ship. Stielo was frustrated with himself, why can't keep his men from dying, he thought. Stielo quickly order Captain Row's ship to flank the lead ship. Stielo then moved his ship into place with the lead ship.

Aboard Talee's Command Ship

" Ma'ma the Resistance Flag ship is move into to flank us". a private said. Talee thought for a second before replying. " Order all ships to fire upon the Flag ship". The private agknoledge the order and carryed it out. After her divsion fired a couple shots most of the flag ship was in pieaces.

Aboard The Ravanger

System failure..System failure...Threes a air lock breach in section 9. The alarms were screaming status reports all over. " Sir we need to evacuate" the blood stained face of a private suggested to Steilo. " Where's Lieutenant Rance?" Stielo snapped. " The private went to go check around the bridge before heading back to Steilo. " Sir he's not on the bridge" the private stated. Maybe he left all ready or maybe he's dead, Stielo quickly thought. Then he issued an order he hated to issue. " Private issue the evacuation order now" Stielo snapped. After the order was issue Stielo was escorted to an escape pod to get off the burning ship.

Aboard Talee's Command Ship

"Ma'ma we have reports on an evacuation, should we shoot them down?" " Negetive Private, stand down the battle is won". she quickly responeded. She was right what was left of the Resistance ship head back into space, and she decide enough blood was spilt for one day so she didn't pursure them. She grab her pad and message General Shaw and Head of State Corriban of there first victory.

Planet: Nyrae Head of State Corriban's office

Corriban was at his desk when he got the report from Talee, he read and sighed with joy. He decided to a toast to celebrate the first victory, so he got up and head to his wine cabinet at the far side of his office. He pulled out his best and went back to his chair, had a couple shots and then notice something by the window. A odd figure, he blinked his eyes then noticed it getting easy to see. He then thought to himself of what it might be, and it brought shivers down his spine. Corriban knew what the figure looked like now. It was the assissan that killed Head of the Union Rylan. Corriban didn't say a word. The hooded figure dropped from the side of Corriban's window in his office and headed toward him. Corriban started shaking in fear. The hooded then spoke. " Head of State Corriban". Corriban was trembling before answering back. " Yes I am him." The hooded figure quickly pulled out his gun to shoot Corriban, when Corriban begin to speak. " At least let me see the face of my killer" he said. The hooded figure didn't move for a couple of seconds, then he toke the hood off so Corriban could see his face. " You?.. It was You?" Those were the last words of Head of State Corriban

www.ingramcontent.com/pod-product-compliance
Lightning Source LLC
Chambersburg PA
CBHW071600170526
45166CB00004B/1735